SUBMARINES and SUBMERSIBLES

Joanne Mattern

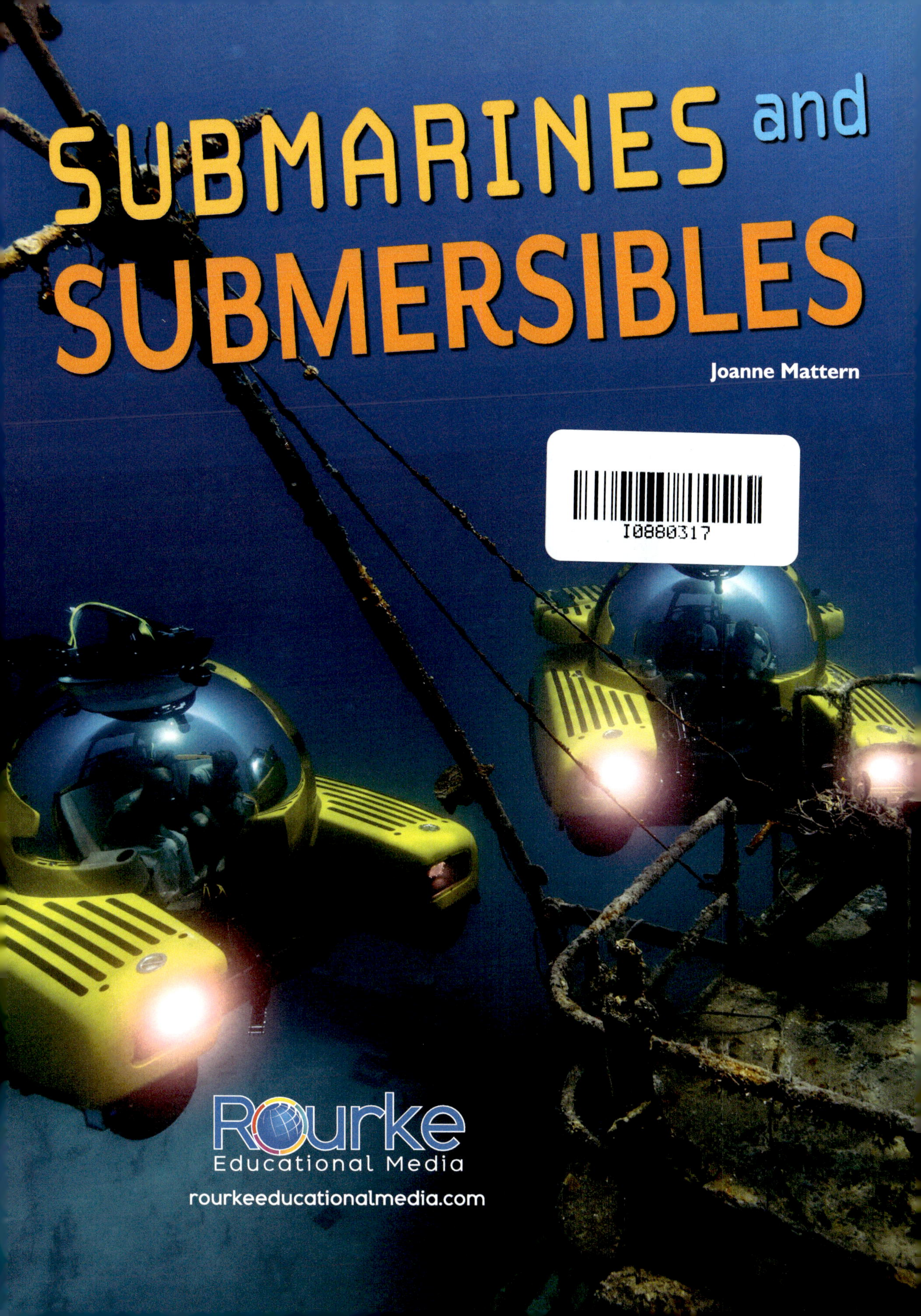

Rourke Educational Media

rourkeeducationalmedia.com

Before, During, and After Reading Activities

Before Reading: Building Background Knowledge and Academic Vocabulary

"Before Reading" strategies activate prior knowledge and set a purpose for reading. Before reading a book, it is important to tap into what your child or students already know about the topic. This will help them develop their vocabulary and increase their reading comprehension.

Questions and activities to build background knowledge:

1. *Look at the cover of the book. What will this book be about?*
2. *What do you already know about the topic?*
3. *Let's study the Table of Contents. What will you learn about in the book's chapters?*
4. *What would you like to learn about this topic? Do you think you might learn about it from this book? Why or why not?*

Building Academic Vocabulary

Building academic vocabulary is critical to understanding subject content.
Assist your child or students to gain meaning of the following vocabulary words.

Content Area Vocabulary

Read the list. What do these words mean?

- ballast tanks
- battery
- compressed
- density
- efficient
- engineers
- mines
- nuclear
- piston
- propellers
- vents
- volume

During Reading: Writing Component

"During Reading" strategies help to make connections, monitor understanding, generate questions, and stay focused.

1. *While reading, write in your reading journal any questions you have or anything you do not understand.*
2. *After completing each chapter, write a summary of the chapter in your reading journal.*
3. *While reading, make connections with the text and write them in your reading journal.*
   a) Text to Self – What does this remind me of in my life? What were my feelings when I read this?
   b) Text to Text – What does this remind me of in another book I've read? How is this different from other books I've read?
   c) Text to World – What does this remind me of in the real world? Have I heard about this before? (News, current events, school, etc....)

After Reading: Comprehension and Extension Activity

"After Reading" strategies provide an opportunity to summarize, question, reflect, discuss, and respond to text. After reading the book, work on the following questions with your child or students to check their level of reading comprehension and content mastery.

1. What are some problems faced by submarine engineers? *(Summarize)*
2. How have submarines been helpful to countries during wars? *(Asking Questions)*
3. Would you like to ride in a submarine? Why or why not? *(Text to Self Connection)*
4. What are some of the most important changes in submarine building? *(Asking Questions)*
5. How might submarines work in the future? *(Asking Questions)*

Extension Activity

Design your own submarine. What kind of engine do you need? How will you design places for people to live? What jobs will your submarine do? How will you keep crew members safe on the submarine?

TABLE OF CONTENTS

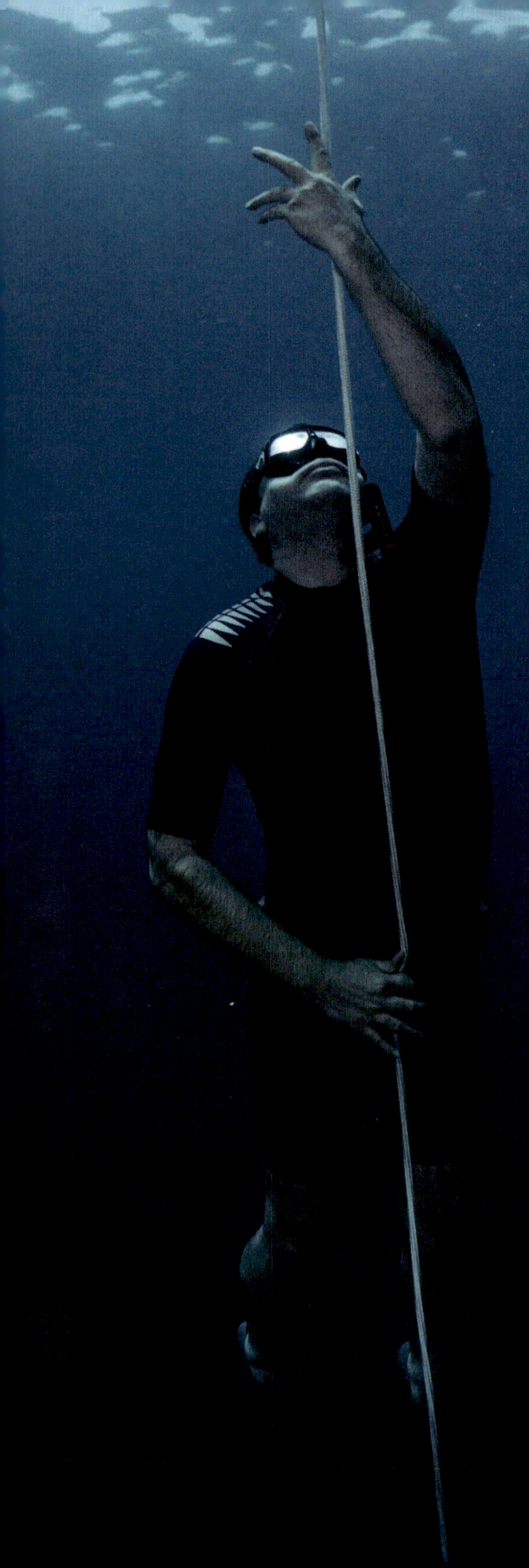

THE FIRST SUBMARINES

Throughout human history, people have been curious about the underwater world. At first, explorers simply held their breath and dove in. But they could only stay submerged for a few minutes before they ran out of air.

Around 3000 BCE, Greek divers used hollow reeds to breathe under water. They also carried bags of air. These did not last long, though. Divers still had to surface quickly.

An Idea Is Bourne

In 1578, William Bourne (1535–1582), an English mathematician and Royal Navy gunner, published a book that described his idea for engineering a submarine. However, Bourne's submarine was never built.

The diving bell gave early explorers another way to breathe under water. These heavy containers were made of wood, metal, or glass. Diving bells sank because the downward force of their weight was greater than the upward push of the water. When the bells sank, a bubble of air was trapped inside. People could use this air to breathe.

Early Bells

Legend has it that Alexander the Great (356–323 BCE) used a type of diving bell in 332 BCE. The modern diving bell was invented in 1690 by Edmund Halley. Halley's diving bell was weighted down with barrels. Tubes brought air to the inside of the bell.

The first working submarine was invented by Corenelis Drebbel in 1620. The *Drebbel* looked like two rowboats stuck on top of each other. People sat inside the *Drebbel* and used oars to row the craft. Drebbel covered his craft with greased leather to keep water out. This method did not work very well.

Drebbel put heavy weights inside his submarine. These weights increased the submarine's **density**, which made it sink. He got rid of the weights to lessen the density and make the submarine rise.

A Trip under the Thames

In 1623, Drebbel's watercraft carried 16 passengers under the River Thames in London. The submarine stayed under water for about three hours, but it did not go very deep.

In 1680, Giovanni Borelli invented **ballast tanks**. Borelli filled goatskin bags inside a submarine with water. This made the submarine denser, so it sank. When he wanted the submarine to rise, Borelli squeezed the water out of the bags.

Giovanni Borelli (1608–1679)

In 1729, Nathaniel Symons invented a sub that could change **volume** to move up and down. A handle inside the submarine pulled the two halves of the ship together. This gave the submarine less volume and density, so it sank. When Symons wanted to return to the surface, he turned the handle the other way. The two halves spread apart. Now the submarine had more volume and was less dense, so it rose to the surface.

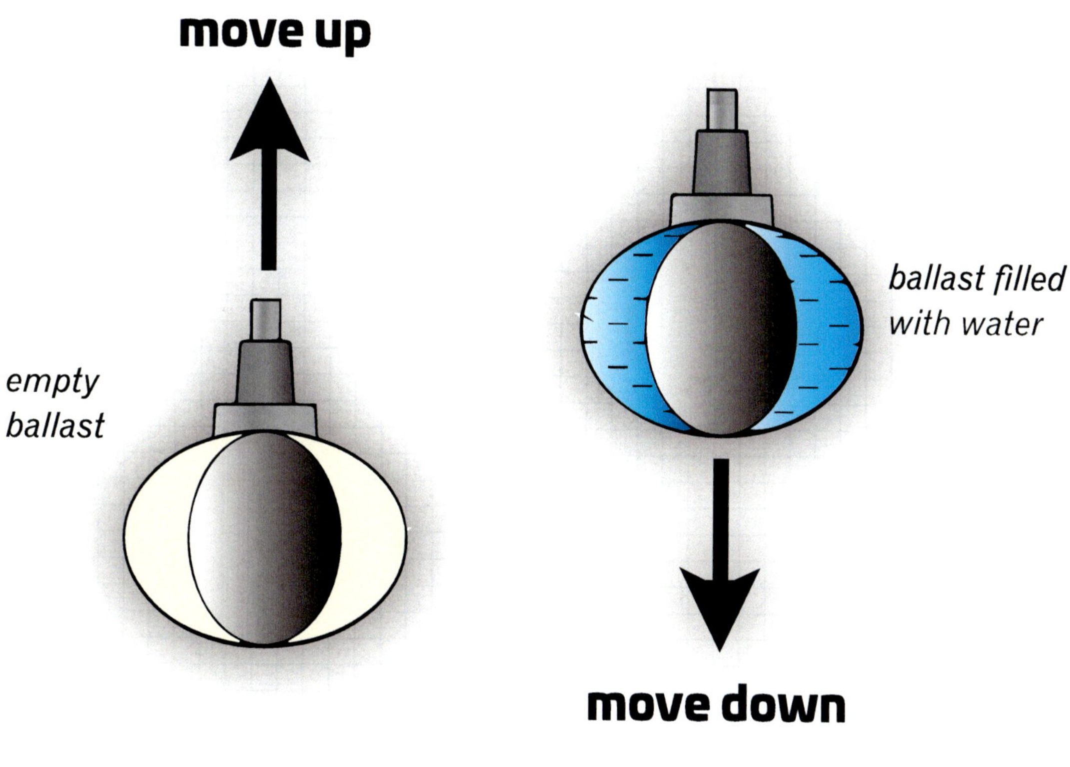

Submarines continued to improve during the late 1800s and early 1900s. A submarine called the *Diver* was invented in 1863. It was the first submarine with an engine.

The engine used **compressed** air. Air was forced from a tank into a tube-shaped cylinder. The moving air pushed a **piston** that turned the **propeller**. The force of the air made the propeller spin very fast.

Ballast tanks attached to the sides of this ship are filled with water to make the boat heavier, so it will sink.

Torpedoes

Torpedoes are underwater missiles that are fired from military submarines. They were invented by Robert Whitehead in 1866. The first torpedoes were metal tubes filled with explosives. Compressed air forced the torpedoes away from the submarine toward their target.

Modern torpedoes are powered by engines, and computers guide them to their targets.

Submarines used steam power in the 1800s. They burned coal to heat water. The hot water produced steam. The steam moved pistons and spun propellers. However, burning coal used up a lot of oxygen. It was not a good fit for submarines.

In 1897, John Philip Holland engineered a submarine that used **battery** power. He called it the *Holland VI*. Batteries created electricity. That electricity then powered a motor to turn the propeller.

John P. Holland peers out of the hatch of the submarine he invented. He was the first to successfully use batteries to power a submarine.

Peeking above the Water

How do sailors see when they are under water? Simon Lake solved that problem by inventing the periscope in 1902. Periscopes are long tubes that have mirrors inside. The mirrors reflect images so sailors can see above the water.

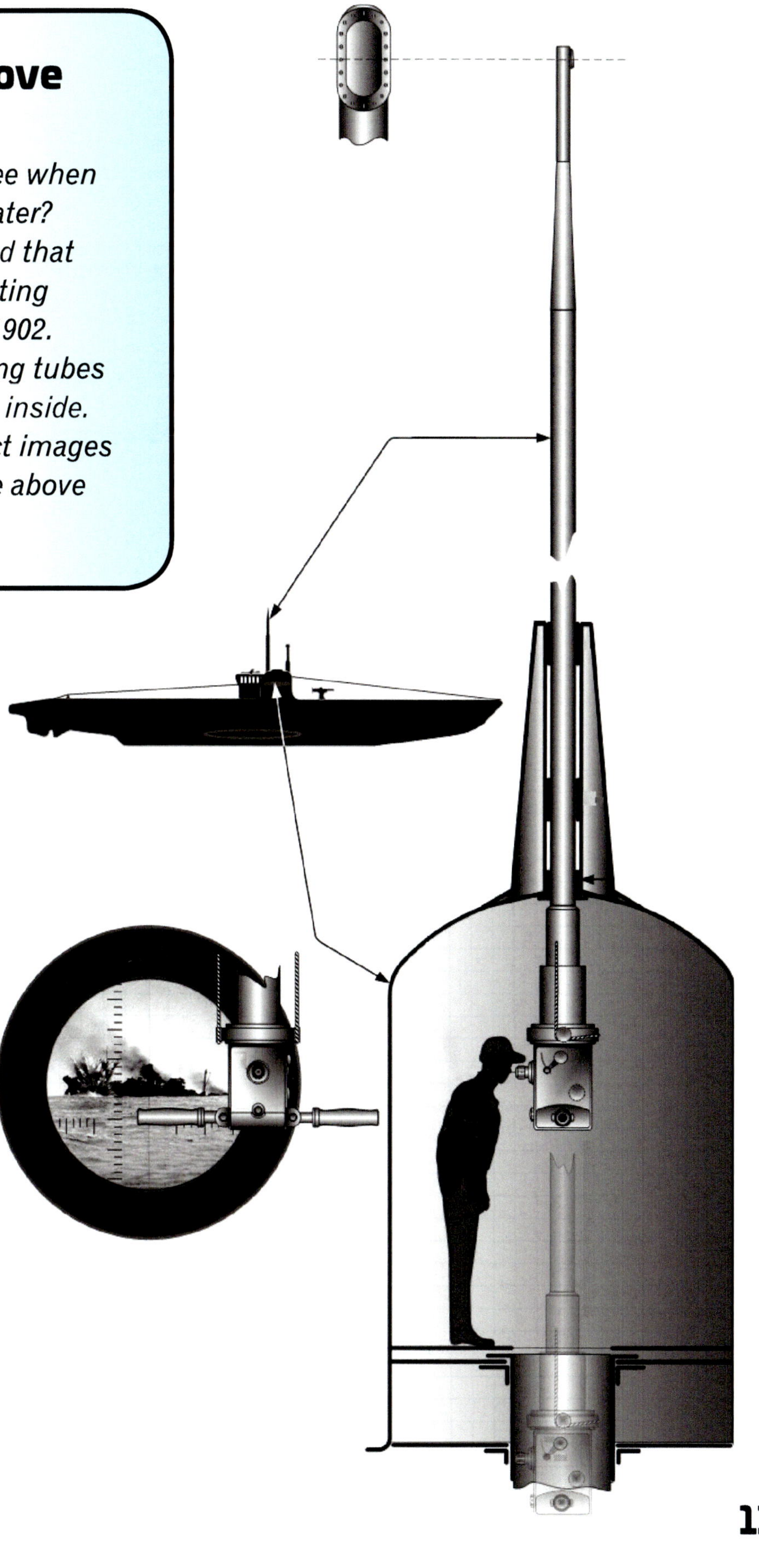

BASICS OF SUBMARINE DESIGN

Military submarine designs are largely confidential, so an enemy can't gain an advantage with the knowledge. But submarines do have some basic standard features.

Submarine Control Surfaces

Submarine Shape

Most submarines are shaped like cigars, wider in the middle and narrow on each end. This shape reduces the drag, or pull, of the water against the submarine when it is submerged. Less drag means the submarine can travel more quickly and easily underwater.

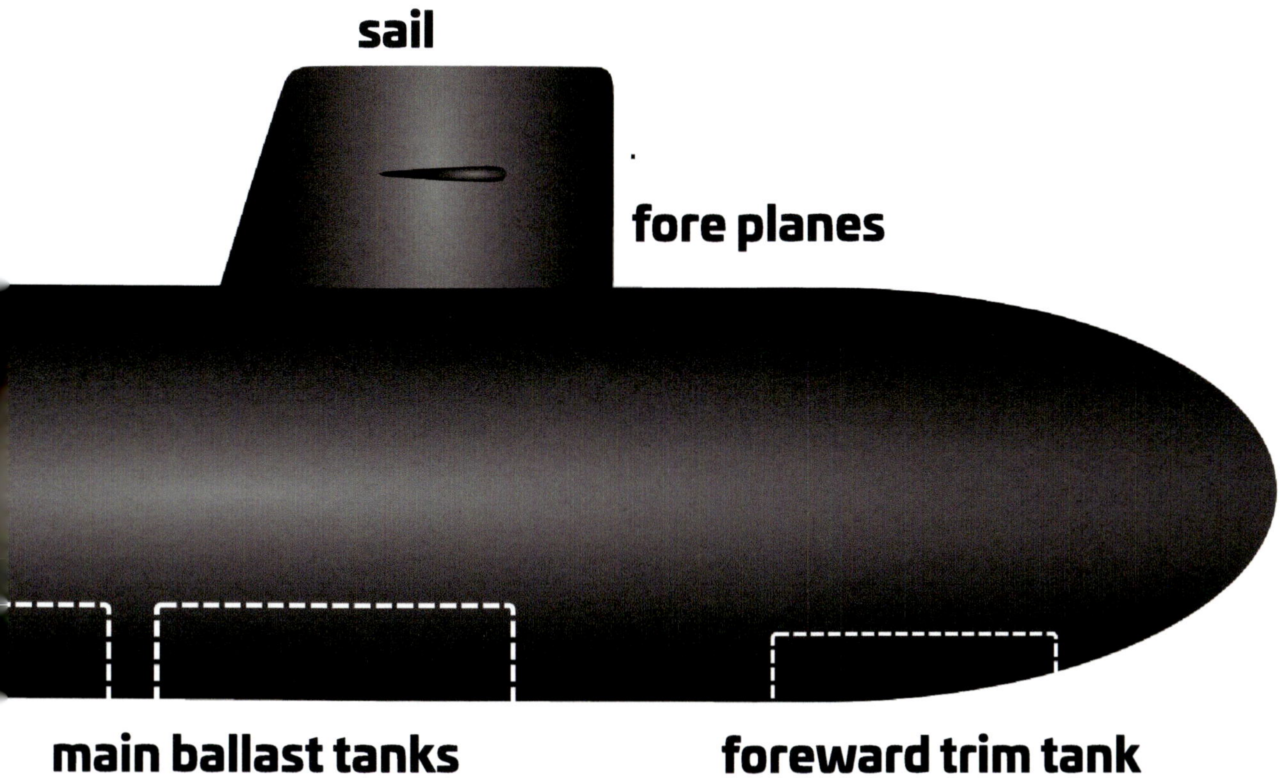

How a Submarine Works

A submarine can float because its weight is equal to the weight of the water it displaces, or shifts. This water displacement creates an upward force called the buoyant force. A regular boat cannot control its buoyancy, but a submarine can!

A submarine's ballast tanks and trim tanks can be filled with water or air. A submarine on the water's surface will have its ballast tanks filled with air. This makes the submarine's density less than the surrounding water. To dive, the ballast tanks are flooded with water to make the submarine sink. This is called negative buoyancy.

A movable set of wing-like structures called tailplanes on the back of the vessel help control the angle of the dive.

The attack submarine USS Columbus *(SSN 762), home ported at Naval Station Pearl Harbor, Hawaii, conducts an emergency surface training exercise 35 miles (56 kilometers) off the coast of Oahu, Hawaii.*

Nuclear submarines use nuclear reactors, steam turbines, and reduction gearing to drive the main propeller shaft. This provides the forward and reverse movement in the water.

Submarines use electric power to operate onboard equipment. Submarines often rely on batteries to supply electrical power to equipment. The batteries are charged by the submarine's diesel engine or nuclear reactor. In an emergency, the batteries may be the submarine's only source of electrical power.

Light does not travel far in the depths of an ocean, which means submarines are essentially traveling blind! To make up for this blindness, submarines are equipped with high-tech navigational equipment. Inertial guidance systems use gyroscopes to track the sub's motion from a fixed starting point. These systems are accurate for up to 150 hours of operation. Then they must be realigned with surface navigation systems such as GPS, radar, or satellite.

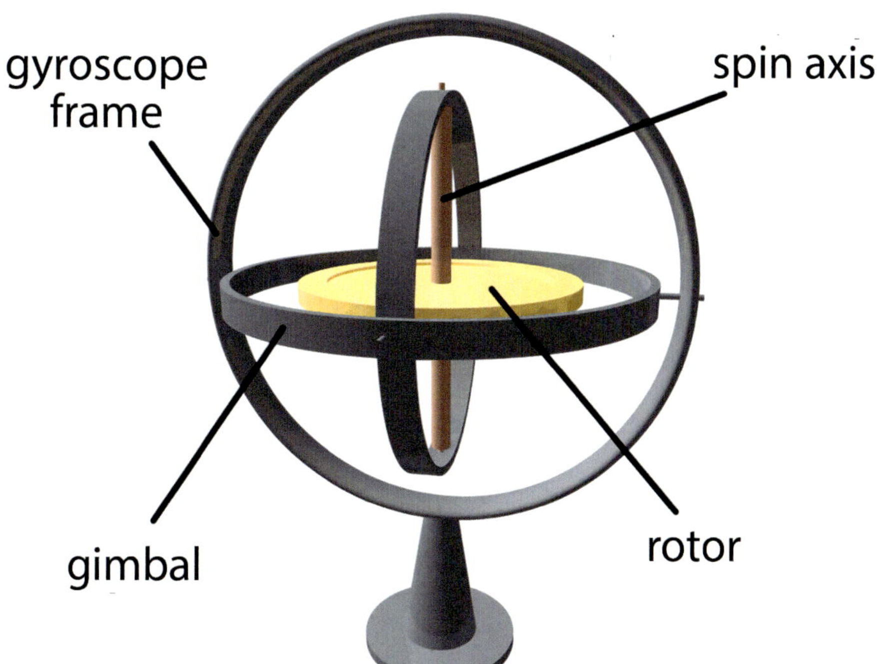

A gyroscope is a wheel or disk mounted to spin rapidly around an axis that is free to turn in various directions.

Submarines use active and passive sonar. Sonar detects the presence and location of submerged objects using sound waves. Active sonar sends out pulses of sound waves that travel through the water, reflect off a target, then return to the ship. Using calculations for the speed of sound under water, computers quickly calculate the distance between the target and the submarine. Passive sonar is the practice of listening for sounds made by objects in the water.

Dolphins use echolocation to find prey, much like a submarine uses active sonar to locate targets.

Did You Know?

The word sonar *was originally an acronym for Sound Navigation And Ranging.*

Blue Nose, Red Nose

U.S. Navy personnel entering the Arctic Circle for the first time are initiated into the Order of the Blue Nose. The Order of the Red Nose is for those who have crossed the Antarctic Circle.

WARTIME SUBMARINES

Submarines have been used in wartime for almost 250 years.

The *Turtle* was the first submarine to attack a war enemy. This submarine was used in the American Revolutionary War (1775–1783). It got its name because it had a curved shape that looked like a turtle's shell.

The plan was for the *Turtle* to drill holes in British ships. Then it would attach **mines** to the ships and blow them up. A sailor inside the ship used handles and pedals to move the submarine's propellers and rudder.

Turtle Attack

On September 6, 1776, an American named Sergeant Lee tried to attack a British ship using the Turtle. Lee drilled a hole in the ship, but he ran out of air before he could attach the mine. Lee had to bring the Turtle to the surface, and the enemy chased him away.

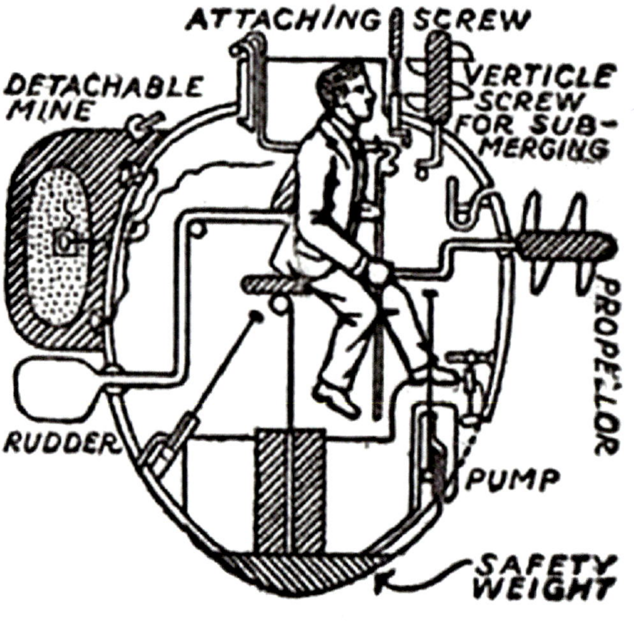

During the American Civil War (1861–1865), the Confederates encouraged the engineering of battle submarines. In 1864, Horace L. Hunley invented a submarine that he named after himself. The *Hunley* could carry a crew of up to nine men. The crew quickly turned a propeller so the submarine could ram enemy ships.

On February 17, 1864, the *Hunley* rammed a Union ship called the U.S.S. *Housatonic*. The *Housatonic* sank. But the *Hunley* also sank. No one knows why.

During World War I (1914–1918), submarines became more popular. Germans called their submarines U-boats, or "undersea boats."

At first, Germany only attacked military ships. Later, Germany attacked merchant ships and passenger liners. In just one day in September 1914, a German U-boat sank three British ships.

Between February and April 1917, U-boats sank more than 500 merchant ships.

German U-boats all followed a similar design. This plan allowed Germany to build a large number of submarines in a short time and also made it simpler to train crew members. By the end of World War I, Germany had more than 350 U-boats in service.

The *Lusitania*

On May 7, 1915, a German U-boat sank a passenger liner called the Lusitania*. Nearly 1,200 people died. One hundred twenty-eight of them were Americans. This attack was a reason why the United States decided to enter World War I.*

Submarines became even more powerful during World War II (1939–1945). New technology such as sonar and radar helped submarines find their targets. Sonar uses sound waves to find submerged objects. Radar uses electromagnetic waves that bounce off objects to show their position.

Sonar operators constantly monitor data on board a nuclear submarine.

A depth charge exploded after it had been dropped from HMS Ceylon. *The ship had just made a call at Colombo, the capital of Sri Lanka, then called Ceylon.*

Fighting Back

As submarines attacked more ships, weapons were developed to stop them. One of the best defenses was the depth charge. Depth charges are explosives that ships drop in the water. An explosion can badly damage a submarine.

USS *Nautilus*

A Long Journey

In 1958, the Nautilus *became the first submarine to travel under the frozen seas of the North Pole. The journey covered about 1,800 miles (3,000 kilometers).*

Submarines used electric power until the 1950s. Then, in 1954, the United States launched the first nuclear submarine. This submarine was called the U.S.S. *Nautilus.*

Nuclear submarines carry a nuclear reactor. Inside the reactor, special elements react with each other to create heat. On the *Nautilus*, that heat powered a steam engine.

Nuclear submarines can travel faster and farther than electric submarines without having to recharge their batteries. The nuclear reactor onboard provides a constant source of power.

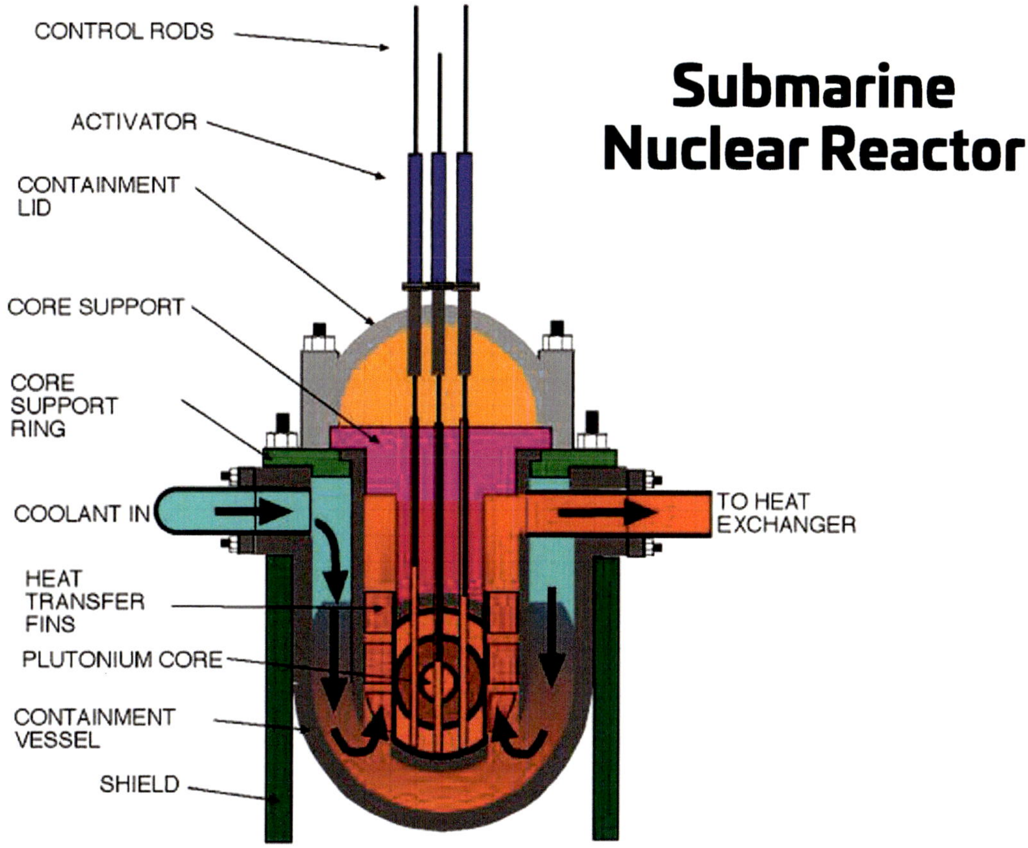

Sailors work in the USS Nautilus' *control room as the sub passes under the ice at the North Pole.*

The Tomahawk Land Attack Missile (TLAM) is a long-range missile used for deep land attack warfare. They are launched from U.S. Navy surface ships and U.S. Navy and United Kingdom Royal Navy submarines.

During the Gulf War (1990–1991), submarines were used to launch land-attack missiles. Before this time, surface ships were the preferred missile launchers because they could carry more. But since submarines could travel stealthily and solo, while ships traveled in groups and were clearly visible on the horizon, submarines were quickly recognized as a preferable missile delivery system.

In the ongoing War on Terror (2001–present), submarines have been called the "perfect platform" for modern warfare.

Two qualities make submarines so effective: stealth and persistence. Submarines can stay concealed under water as their crews collect intelligence, rising to periscope depth to take pictures and listen to electronic transmissions.

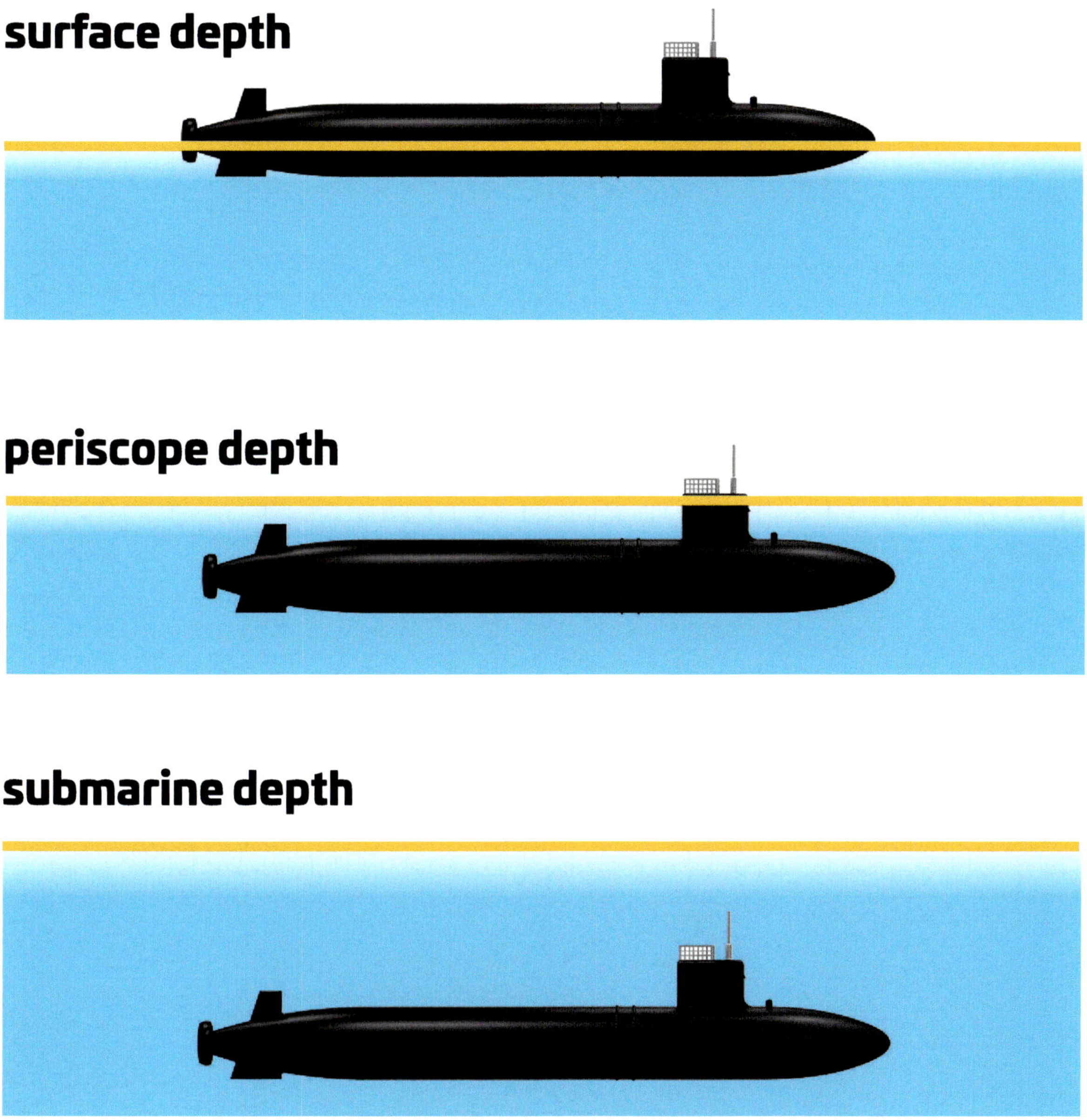

While airplanes and satellites pass over targets, submarines can remain stationed for weeks or months.

Fast attack submarines are armed with 12 cruise missiles with conventional warheads, in addition to traditional torpedoes. They can also land special operations teams to conduct raids and collect intelligence.

The U.S. submarine force is used for:

- *Peacetime Engagement*
- *Surveillance*
- *Special Operations*
- *Precision Strike*
- *Battlegroup Operations*
- *Sea Denial*

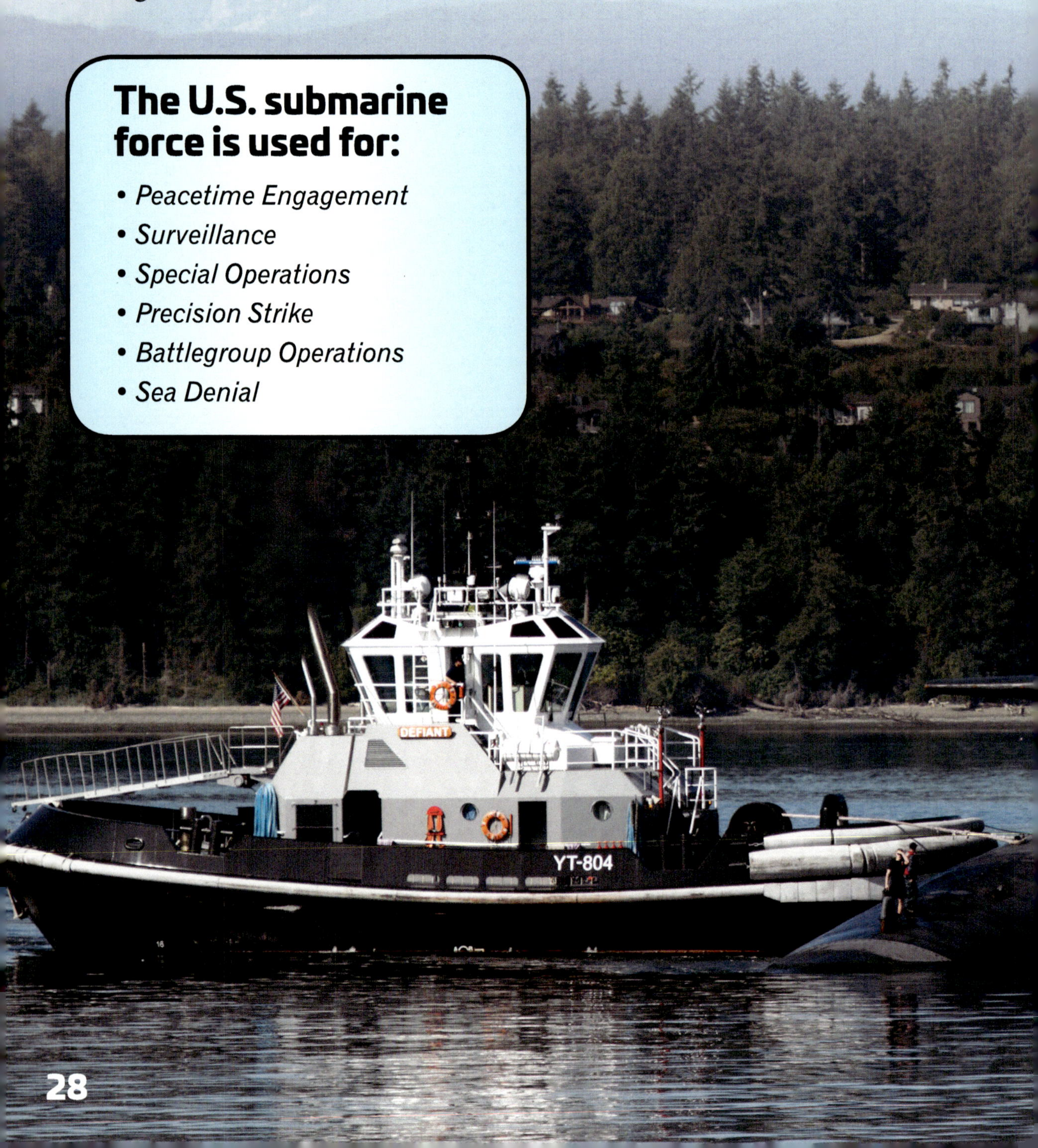

Peacetime

During peacetime, a submarine deployment can be used to show interest in a region. Authorities may also choose to deploy a submarine covertly, keeping the submarine submerged and out of sight until the time is right to reveal its presence. A nuclear submarine's high speed and lengthy endurance make it ideal for rapid deployment when needed.

Sailors stand on top of the USS Michigan *as it cruises into its home port in Washington state. The sub returned from a 20-month tour of the waters around Guam, where it performed surveillance and training operations.*

Surveillance

Submarines can enter an area to listen, watch, and collect information without being seen. Unlike satellites and aircraft, which can be inhibited by a target's location, weather conditions, and cloud cover, submarines can get closer to the action and capture signals that these surveillance methods may not.

A member of SEAL Delivery Vehicle Team climbs aboard one of the team's SEAL Delivery Vehicles, or SDVs, before launching it from the back of a submarine. SDVs carry Navy SEALs from a submarine to enemy targets, allowing them to stay underwater and undetected.

Special Operations

Submarines are used to carry special operations units such as commandos and agents on high-risk missions. Most U.S. special operations using submarines are performed by SEALs, Sea-Air-Land teams trained for missions behind enemy lines. These missions include search and rescue, intelligence gathering, and sabotage.

A member of a SEAL Delivery Vehicle team prepares to launch an SDV from the USS Philadelphia *during a training exercise.*

Sea Denial

Submarines can be used to block enemy ships and subs from using the oceans and seas. They can block enemy ports and sink military and merchant ships, reducing an enemy's fleet and cutting off supplies.

Precision Strike

U.S. attack submarines carry Tomahawk Land Attack Missiles (TLAM). These missiles allow for long-range, precision strikes with conventional warheads against land targets.

Is this the submarine of the future? Four U.S. strategic missile submarines will be transformed into SSGN or Tactical Trident submarines. The SSGNs will support and launch up to 154 Tomahawk missiles. The 22 missile tubes will also provide the capability to carry other payloads, such as unmanned underwater and aerial vehicles and special forces equipment.

Battlegroup Operations

A battlegroup is a military force made up of several types of troops that work together. Two attack submarines are typically assigned to each U.S. Navy battle group. These submarines join in all pre-deployment operational training and exercises. During this time, submarine command is often shifted to amphibious group commanders, battle group commanders, or NATO commanders.

Finding the *Titanic*

In 1985, scientists finally found the wreck of the Titanic *deep under the Atlantic Ocean. ROVs, or remote-controlled submarines, explored the wreck and took pictures.*

RESEARCH AND RECREATIONAL SUBS

Modern submarines aren't only used by the military. They are also used for research and recreation.

Submarines are important tools for scientists. These underwater ships can go much deeper than people can. Explorers use them to discover underwater wrecks. These shipwrecks contain objects that teach us about our history. Exploring a shipwreck can also solve mysteries, such as why the ship sank or if it carried valuable cargo such as gold or jewels.

Submersibles

Smaller vehicles, called submersibles, can go even farther. A submersible is smaller than a submarine. This allows submersibles to travel into places submarines can't go. While a submarine can travel long distances under its own power, a submersible does not have a large source of power. It usually runs on batteries.

Submersibles can travel deep under the ocean. They can explore caves and underwater **vents**. Submersibles have discovered ecosystems in places where scientists did not think anything could live.

Deep Submergence Rescue Vehicles

What if there is an emergency? Deep Submergence Rescue Vehicles (DSRVs) attach to a submarine's escape hatch so trapped sailors can climb out. Remote Operated Vehicles (ROVs) can also work to untangle submarines if they get caught in wires or nets.

PERSONAL SUBMARINES

Seabreacher X

Specifications:

Length: 17 feet (5.08 meters)
Width: 3.5 feet (1.0 meters)
Wingspan: 6 feet, 6 inches (1.9 meters)
Approximate Weight: 1,450 pounds (657.7 kilograms)
Maximum Speed: 50 miles (80.5 kilometers) per hour on surface, 25 miles (40 kilometers) underwater
Maximum Depth: 6 feet (1.83 meters) for brief periods

The Seabreacher X, offered by Innespace Productions, is a personal recreational submarine. The two-seater sub features a shark-style body and a thrust system that mimics the tail movements of dolphins and sharks. And it can jump 90 degrees out of the water!

Seabreacher Y

Specifications:

Length: 190 inches (4.83 meters)

Width: 42 inches (1.07 meters)

Wingspan: 78 inches (1.98 meters)

Height: 58 inches (1.47 meters)

Approximate weight: 1,450 pounds (657.71 kilometers)

Just like the sea creature it is named for, the Orca's design features a large tail, pectoral fins, and a blowhole. The Orca also has a supercharged engine, and its clear acrylic canopy gives passengers an amazing view.

A Submarine Racing Contest

Every other year, engineering students from around the world take part in the European International Submarine Race. Teams design and build submarines and race them underwater. The contest helps ***engineers*** *learn new ways to build submarines. It could also lead to exciting new technology!*

Seabreacher Z

Specifications:

Length: 192–200 inches (4.88m–5.08 meters)
Width: 42 inches (1.06 meters)
Wingspan: 78 inches (1.98 meters)
Height: 45 inches (1.14 meters)
*Snorkel Down
Approximate weight: 1,450 pounds (657.71 kilograms)
Fuel capacity: 14 gallons (53 liters)

Seabreacher's Z model has a dolphin-style body. A special snorkel feature allows passengers to experience thrilling high-speed barrel rolls, and its supercharged engine makes it the most extreme personal watercraft on the market today.

www.seabreacher.com

Seabreacher Sailfish

The Seabreacher Sailfish has a powerful engine and a removable sail/snorkel that allows 360-degree barrel rolls on the water. This amazing sub also has a water cannon that fires from the nose of the craft. A fully loaded model costs about 125 thousand dollars.

SUBMARINES OF THE FUTURE

What does the future of submarine design look like? Marine engineers are developing new concepts every day.

Marine engineers and architects usually start by working on computers. They use math formulas and computer programs to design submarines. Engineers can plan and design new features and test them on the computer. They may also travel out to sea on a submarine to see if the craft is working properly.

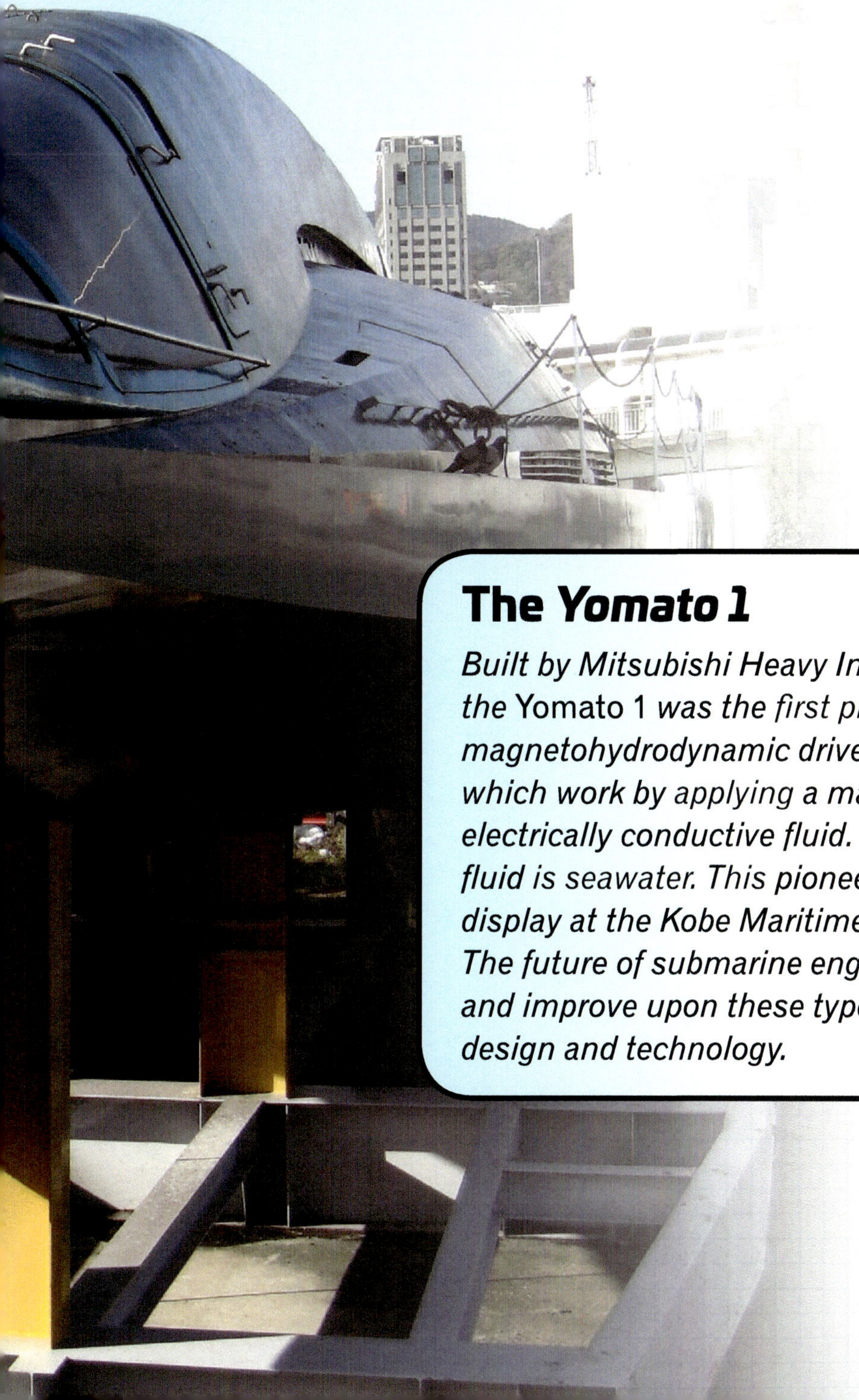

The *Yomato 1*

Built by Mitsubishi Heavy Industries, Ltd., the Yomato 1 *was the first prototype to use magnetohydrodynamic drives (MHDs), which work by applying a magnetic field to an electrically conductive fluid. In this case, that fluid is seawater. This pioneering vessel is now on display at the Kobe Maritime Museum in Japan. The future of submarine engineering will rely and improve upon these types of innovations in design and technology.*

Engineers are working on many new technologies. For example, future submarines probably won't have propellers. They may move by forcing jets of water to push them where the crew wants to go. Or they might use magnets to steer. Engineers are always looking for ways to make submarines faster and more **efficient**.

Goodbye, Periscope?

New submarines do not have periscopes like earlier submarines do. Instead, a fiber optic imaging mast sits in the sail outside the hull and sends electronic images to the crew on the sub. Getting rid of the periscope makes the hull of the ship stronger because it does not need to have an opening for the periscope to poke through.

Engineers are looking for ways to make submarines quieter. They want submarines to be able to travel to enemy ships without the enemy detecting their presence. To do this, submarines might become like underwater aircraft carriers. Subs will "park" underwater and release submersibles or drones. These drones could spy on or attack the enemy without anyone knowing they are there until it is too late. This is just one idea that could change the role of submarines in a big way.

TIMELINE

1578: William Bourne publishes his idea for making a submarine.

1620: The *Drebbel* is launched.

1680: Giovanni Borelli invents ballast tanks.

1729: Nathaniel Symons invents a submarine that can adjust its volume.

1776: David Bushnell invents the *Turtle* to attack British ships.

1797: Robert Fulton invents the slow-moving *Nautilus*.

1863: Charles Burn and Simon Bourgeois invent the *Diver*, the first submarine with an engine.

1864: The *H.L. Hunley* successfully attacks the U.S.S. *Housatonic*, but the *Hunley* is lost at sea.

1866: Robert Whitehead invents the torpedo.

1902: Simon Lake invents the periscope.

1953: The first ROV is invented.

1954: The U.S.S. *Nautilus* is the first nuclear-powered submarine.

1986: The ROV *Jason Jr.* photographs the inside of the *Titanic* wreck.

GLOSSARY

ballast tanks (BAL-uhst tanks): large tanks in a submarine that are filled with water to make the submarine sink, or with air to make it rise to the surface

battery (BAT-uh-ree): a container filled with chemicals that produces electrical power

compressed (kuhm-PRESSED): squeezed into a small space

density (DEN-si-tee): how heavy or light something is

efficient (uh-FISH-uhnt): not wasting time or energy

engineers (en-juh-NEERZ): people who are trained to design and build machines or structures

mines (mines): bombs placed underground or underwater

nuclear (NOO-klee-ur): energy created by splitting atoms

piston (PISS-tuhn): a disk or rod that moves back and forth in a large cylinder

propeller (pruh-PEL-ur): a device with rotating blades that provide force to move a vehicle

vents (vents): openings through which smoke or heat can escape

volume (VOL-yuhm): the amount of space taken up by a three-dimensional shape, such as a boat

INDEX

SHOW WHAT YOU KNOW

1. Why did people want to invent a way to travel under water?
2. What are some problems facing the development of submarines?
3. How have submarines been used during wartime?
4. What technologies have changed the way submarines work and what they can do?
5. What are some ways submarines are useful to people now and in the future?

FURTHER READING

Mahaney, Ian F., *Extreme Submarines*, PowerKids Press, 2015.

Spilsbury, Louise A., Spilsbury, Richard, *Submarines at Sea*, PowerKids Press, 2017.

West, David, *Submarines*, Smart Apple Media, 2017.

ABOUT THE AUTHOR

Joanne Mattern is the author of numerous nonfiction books for children. She especially enjoys writing about science and history topics. Her titles for Rourke include *Floods* (Devastating Disasters series), *Bridges*, and *Tunnels* (Engineering Wonders series). Joanne lives in New York with her husband, four children, and several pets.

Meet The Author!
www.meetREMauthors.com

www.rourkeeducationalmedia.com

PHOTO CREDITS: Cover & Title Pg ©Nick Verola Triton (Shamin Abas); Pg 3 ©Petty Officer 2nd Class Amanda G/U.S. Navy; Pg 4 ©mihtiander; Pg 5 ©Chronicle / Alamy Stock Photo; Pg 6 ©John Cairns / Alamy Stock Photo; Pg 7 ©akinshin, Science History Images / Alamy Stock Photo; Pg 8 ©Historical Images Archive / Alamy Stock Photo; Pg 9 ©iv-serg; Pg 10 ©Everett Collection Historical / Alamy Stock Photo; Pg 11 ©Tamasflex, Pg 13 ©Wiki; Pg 14 ©David C. Duncan; Pg 15 ©Wiki; Pg 16 ©marshalgonz; Pg 17 ©Wiki; Pg 18 ©Wiki; Pg 19 ©Wiki; Pg 20 ©Library of Congress, ©Wiki; Pg 21 ©Wiki; Pg 22 ©Buzz Pictures / Alamy Stock Photo; Pg 23 ©Wiki; Pg 24 ©U.S. Navy; Pg 25 ©Wiki, ©Keystone Pictures USA / Alamy Stock Photo; Pg 26 ©Kenneth Moll; Pg 29 ©Mass Communication Specialist 2n; Pg 30 & 31©PHC Andrew McKaskle; Pg 33 ©U.S. Navy; Pg 34 ©Wiki; Pg 36-39 ©Wiki; Pg 40 ©Wiki; Pg 42 ©MR1805; Pg 44 ©MR1805

Edited by: Keli Sipperley
Cover design by: Rhea Magaro-Wallace
Interior design by: Kathy Walsh

Library of Congress PCN Data

Submarines and Submersibles / Joanne Mattern
(Engineering Wonders)
ISBN 978-1-64369-050-6 (hard cover)
ISBN 978-1-64369-088-9 (soft cover)
ISBN 978-1-64369-197-8 (e-Book)
Library of Congress Control Number: 2018956028

Rourke Educational Media
Printed in the United States of America,
North Mankato, Minnesota